500 SAMURAI
PUZZLE BOOK FOR ADULTS

EASY TO HARD

I0491854

Samurai Sudoku Logic Puzzles Book is a collection of 500 easy to hard Sudoku puzzles, the book is designed to help keep your brain cognitively fit, flexible, and young.

Elmsleigh Designs

CONTENTS

1# EASY

2# EASY

3# EASY

1

4# EASY

SOLUTION ON PAGE 126

| PUZZLE #501 | PUZZLE #502 | PUZZLE #503 | PUZZLE #504 |

1# EASY

6	10		7	24		21		11
	19			5				
14		10	9		4		12	
	8			17	8	12		13
	11					12		
10		15			12			
15		8	12	9		16		
9				9	11		5	11
	22				8			

2# EASY

7		12		11	3		18	14
7		13						10
15		9	20					7
3			11		20			
8	17		5		13	14		
	14	7	17	15		8	20	
18								9
	6			9		12		
	11			11			11	

3# EASY

16	8		13		9	24		
	7		8				15	
	14			15			6	
13	13			10	9		5	8
	13		6		13	14		
9		19		19			13	8
14					16			
11			6			14		11
4			13			9		

4# EASY

3		18	12			15	13	
19	11		8	19			3	
				15			14	
	10	10			12		7	
		21		5	9		11	
17		14			6		22	
21			15	23				
	9	4					14	
			10			15		

1

SOLUTION ON PAGE 126

5# EASY

6# EASY

2

7# EASY

8# EASY

SOLUTION ON PAGE 125

PUZZLE #497

PUZZLE #498

PUZZLE #499

PUZZLE #500

9# EASY

10# EASY

11# EASY

12# EASY

SOLUTION ON PAGE 124

PUZZLE #493   PUZZLE #494   PUZZLE #495   PUZZLE #496

13# EASY

14# EASY

4

15# EASY

16# EASY

SOLUTION ON PAGE 123

PUZZLE #489 PUZZLE #490 PUZZLE #491 PUZZLE #492

17# EASY

18# EASY

5

19# EASY

20# EASY

SOLUTION ON PAGE 122

21# EASY

22# EASY

23# EASY

24# EASY

SOLUTION ON PAGE 121

PUZZLE #481

PUZZLE #482

PUZZLE #483

PUZZLE #484

25# EASY

26# EASY

27# EASY

28# EASY

SOLUTION ON PAGE 120

PUZZLE #477   **PUZZLE #478**   **PUZZLE #479**   **PUZZLE #480**

29# EASY

30# EASY

8

31# EASY

32# EASY

SOLUTION ON PAGE 119

SOLUTION ON PAGE 119

| PUZZLE #473 | PUZZLE #474 | PUZZLE #475 | PUZZLE #476 |

33# EASY

34# EASY

35# EASY

9

36# EASY

SOLUTION ON PAGE 118

PUZZLE #469

PUZZLE #470

PUZZLE #471

PUZZLE #472

37# EASY

38# EASY

10

39# EASY

40# EASY

SOLUTION ON PAGE 117

PUZZLE #465 PUZZLE #466 PUZZLE #467 PUZZLE #468

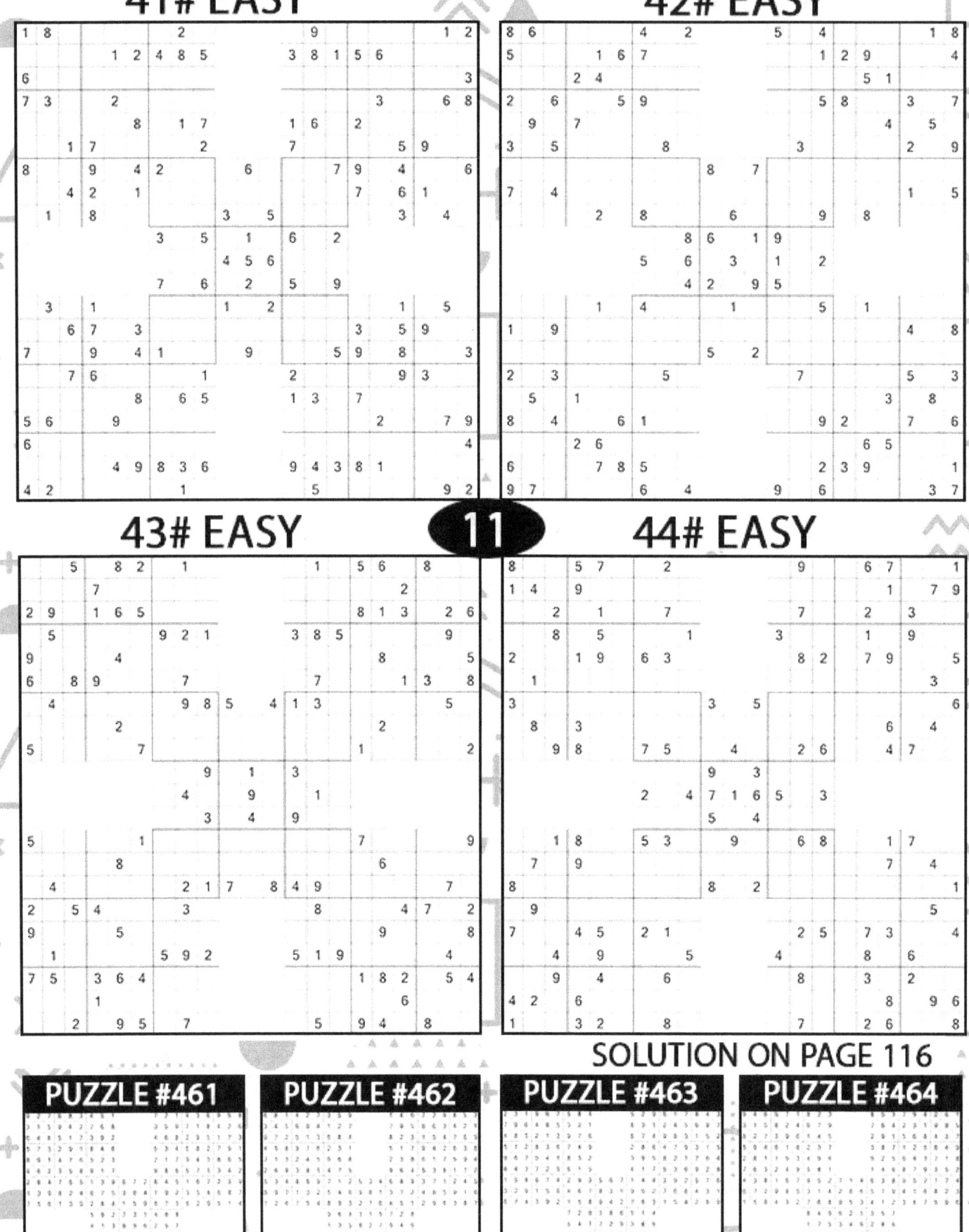

41# EASY

42# EASY

43# EASY

11

44# EASY

SOLUTION ON PAGE 116

PUZZLE #461

PUZZLE #462

PUZZLE #463

PUZZLE #464

45# EASY

46# EASY

47# EASY

12

48# EASY

SOLUTION ON PAGE 115

| PUZZLE #457 | PUZZLE #458 | PUZZLE #459 | PUZZLE #460 |

49# EASY

50# EASY

51# EASY

52# EASY

SOLUTION ON PAGE 114

PUZZLE #453   PUZZLE #454   PUZZLE #455   PUZZLE #456

53# MEDIUM

54# MEDIUM

14

55# MEDIUM

56# MEDIUM

SOLUTION ON PAGE 113

57# MEDIUM

58# MEDIUM

59# MEDIUM

60# MEDIUM

SOLUTION ON PAGE 112

PUZZLE #445   PUZZLE #446   PUZZLE #447   PUZZLE #448

61# MEDIUM

62# MEDIUM

63# MEDIUM

16

64# MEDIUM

SOLUTION ON PAGE 111

65# MEDIUM

66# MEDIUM

17

67# MEDIUM

68# MEDIUM

SOLUTION ON PAGE 110

PUZZLE #437   PUZZLE #438   PUZZLE #439   PUZZLE #440

69# MEDIUM

70# MEDIUM

18

71# MEDIUM

72# MEDIUM

SOLUTION ON PAGE 109

SOLUTION ON PAGE 109

PUZZLE #433 **PUZZLE #434** **PUZZLE #435** **PUZZLE #436**

73# MEDIUM

74# MEDIUM

19

75# MEDIUM

76# MEDIUM

SOLUTION ON PAGE 108

PUZZLE #429 **PUZZLE #430** **PUZZLE #431** **PUZZLE #432**

77# MEDIUM

78# MEDIUM

79# MEDIUM

20

80# MEDIUM

SOLUTION ON PAGE 107

SOLUTION ON PAGE 107

PUZZLE #425	PUZZLE #426	PUZZLE #427	PUZZLE #428

81# MEDIUM

82# MEDIUM

21

83# MEDIUM

84# MEDIUM

SOLUTION ON PAGE 106

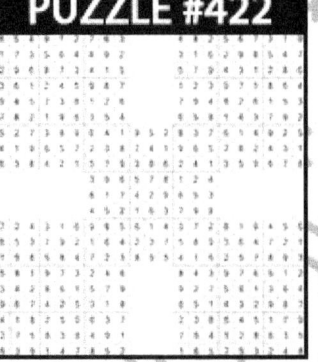

PUZZLE #421 PUZZLE #422 PUZZLE #423 PUZZLE #424

85# MEDIUM

86# MEDIUM

87# MEDIUM

88# MEDIUM

SOLUTION ON PAGE 105

PUZZLE #417

PUZZLE #418

PUZZLE #419

PUZZLE #420

89# MEDIUM

90# MEDIUM

91# MEDIUM

23

92# MEDIUM

SOLUTION ON PAGE 104

PUZZLE #413

PUZZLE #414

PUZZLE #415

PUZZLE #416

93# MEDIUM

94# MEDIUM

95# MEDIUM

24

96# MEDIUM

SOLUTION ON PAGE 103

PUZZLE #409

PUZZLE #410

PUZZLE #411

PUZZLE #412

97# MEDIUM

98# MEDIUM

99# MEDIUM

25

100# MEDIUM

SOLUTION ON PAGE 102

PUZZLE #405 | PUZZLE #406 | PUZZLE #407 | PUZZLE #408

101# MEDIUM

102# MEDIUM

103# MEDIUM

26

104# MEDIUM

SOLUTION ON PAGE 101

PUZZLE #401

PUZZLE #402

PUZZLE #403

PUZZLE #404

105# MEDIUM

106# MEDIUM

27

107# MEDIUM

108# MEDIUM

SOLUTION ON PAGE 100

PUZZLE #397

PUZZLE #398

PUZZLE #399

PUZZLE #400

109# MEDIUM

110# MEDIUM

111# MEDIUM

112# MEDIUM

SOLUTION ON PAGE 99

PUZZLE #393

PUZZLE #394

PUZZLE #395

PUZZLE #396

113# MEDIUM

114# MEDIUM

29

115# MEDIUM

116# MEDIUM

SOLUTION ON PAGE 98

PUZZLE #389

PUZZLE #390

PUZZLE #391

PUZZLE #392

117# MEDIUM

118# MEDIUM

119# MEDIUM

30

120# MEDIUM

SOLUTION ON PAGE 97

SOLUTION ON PAGE 97

PUZZLE #385 PUZZLE #386 PUZZLE #387 PUZZLE #388

121# MEDIUM

122# MEDIUM

123# MEDIUM

124# MEDIUM

SOLUTION ON PAGE 96

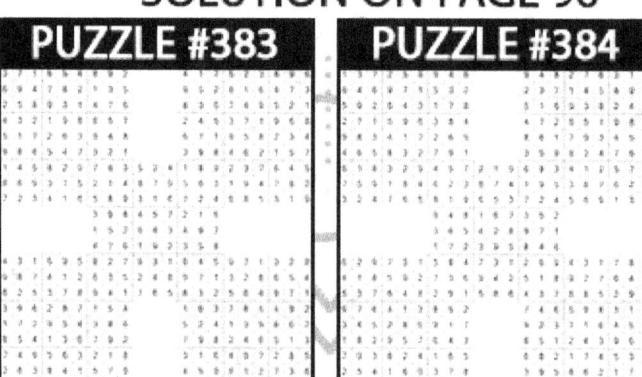

PUZZLE #381 PUZZLE #382 PUZZLE #383 PUZZLE #384

125# MEDIUM

126# MEDIUM

32

127# MEDIUM

128# MEDIUM

SOLUTION ON PAGE 95

PUZZLE #377

PUZZLE #378

PUZZLE #379

PUZZLE #380

129# MEDIUM

130# MEDIUM

131# MEDIUM

132# MEDIUM

SOLUTION ON PAGE 94

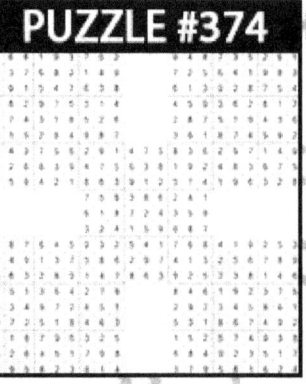

PUZZLE #373   **PUZZLE #374**   **PUZZLE #375**   **PUZZLE #376**

133# MEDIUM

134# MEDIUM

135# MEDIUM

34

136# MEDIUM

SOLUTION ON PAGE 93

PUZZLE #369

PUZZLE #370

PUZZLE #371

PUZZLE #372

137# MEDIUM

138# MEDIUM

35

139# MEDIUM

140# MEDIUM

SOLUTION ON PAGE 92

| PUZZLE #365 | PUZZLE #366 | PUZZLE #367 | PUZZLE #368 |

141# MEDIUM

142# MEDIUM

143# MEDIUM

36

144# MEDIUM

SOLUTION ON PAGE 91

PUZZLE #361

PUZZLE #362

PUZZLE #363

PUZZLE #364

145# MEDIUM

146# MEDIUM

37

147# MEDIUM

148# MEDIUM

SOLUTION ON PAGE 90

149# MEDIUM

150# MEDIUM

38

151# MEDIUM

152# MEDIUM

SOLUTION ON PAGE 89

SOLUTION ON PAGE 89

| PUZZLE #353 | PUZZLE #354 | PUZZLE #355 | PUZZLE #356 |

153# MEDIUM

154# MEDIUM

39

155# MEDIUM

156# MEDIUM

SOLUTION ON PAGE 88

PUZZLE #349

PUZZLE #350

PUZZLE #351

PUZZLE #352

157# MEDIUM

158# MEDIUM

159# MEDIUM

40

160# MEDIUM

SOLUTION ON PAGE 87

PUZZLE #345   PUZZLE #346   PUZZLE #347   PUZZLE #348

161# MEDIUM

162# MEDIUM

163# MEDIUM

41

164# MEDIUM

SOLUTION ON PAGE 86

SOLUTION ON PAGE 86

PUZZLE #341

PUZZLE #342

PUZZLE #343

PUZZLE #344

165# MEDIUM

166# MEDIUM

42

167# MEDIUM

168# MEDIUM

SOLUTION ON PAGE 85

PUZZLE #337

PUZZLE #338

PUZZLE #339

PUZZLE #340

169# MEDIUM

170# MEDIUM

171# MEDIUM

43

172# MEDIUM

SOLUTION ON PAGE 84

PUZZLE #333

PUZZLE #334

PUZZLE #335

PUZZLE #336

173# MEDIUM

174# MEDIUM

175# MEDIUM

44

176# MEDIUM

SOLUTION ON PAGE 83

177# MEDIUM

178# MEDIUM

45

179# MEDIUM

180# MEDIUM

SOLUTION ON PAGE 82

PUZZLE #325 **PUZZLE #326** **PUZZLE #327** **PUZZLE #328**

181# MEDIUM

182# MEDIUM

183# MEDIUM

46

184# MEDIUM

SOLUTION ON PAGE 81

185# MEDIUM

186# MEDIUM

187# MEDIUM

47

188# MEDIUM

SOLUTION ON PAGE 80

| PUZZLE #317 | PUZZLE #318 | PUZZLE #319 | PUZZLE #320 |

189# MEDIUM

190# MEDIUM

191# MEDIUM

48

192# MEDIUM

SOLUTION ON PAGE 79

PUZZLE #313

PUZZLE #314

PUZZLE #315

PUZZLE #316

193# MEDIUM

194# MEDIUM

195# MEDIUM

49

196# MEDIUM

SOLUTION ON PAGE 78

PUZZLE #309

PUZZLE #310

PUZZLE #311

PUZZLE #312

197# MEDIUM

198# MEDIUM

199# MEDIUM

50

200# MEDIUM

SOLUTION ON PAGE 77

PUZZLE #305

PUZZLE #306

PUZZLE #307

PUZZLE #308

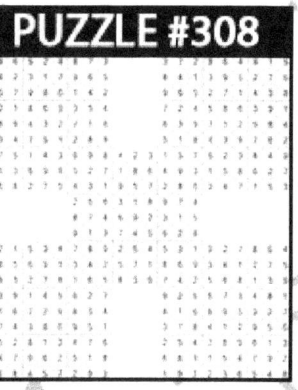

201# MEDIUM

202# MEDIUM

51

203# MEDIUM

204# MEDIUM

SOLUTION ON PAGE 76

205# MEDIUM

206# MEDIUM

207# MEDIUM

52

208# MEDIUM

SOLUTION ON PAGE 75

PUZZLE #297

PUZZLE #298

PUZZLE #299

PUZZLE #300

209# MEDIUM

210# MEDIUM

53

211# MEDIUM

212# MEDIUM

SOLUTION ON PAGE 74

SOLUTION ON PAGE 74

PUZZLE #293

PUZZLE #294

PUZZLE #295

PUZZLE #296

213# MEDIUM

214# MEDIUM

215# MEDIUM

54

216# MEDIUM

SOLUTION ON PAGE 73

PUZZLE #289

PUZZLE #290

PUZZLE #291

PUZZLE #292

217# MEDIUM

218# MEDIUM

55

219# MEDIUM

220# MEDIUM

SOLUTION ON PAGE 72

PUZZLE #285

PUZZLE #286

PUZZLE #287

PUZZLE #288

221# MEDIUM

222# MEDIUM

223# MEDIUM

56

224# MEDIUM

SOLUTION ON PAGE 71

SOLUTION ON PAGE 71

PUZZLE #281 **PUZZLE #282** **PUZZLE #283** **PUZZLE #284**

225# MEDIUM

226# MEDIUM

227# MEDIUM

57

228# MEDIUM

SOLUTION ON PAGE 70

PUZZLE #277

PUZZLE #278

PUZZLE #279

PUZZLE #280

229# MEDIUM

230# MEDIUM

58

231# MEDIUM

232# MEDIUM

SOLUTION ON PAGE 69

PUZZLE #273

PUZZLE #274

PUZZLE #275

PUZZLE #276

233# MEDIUM

234# MEDIUM

235# MEDIUM

59

236# MEDIUM

SOLUTION ON PAGE 68

SOLUTION ON PAGE 68

| PUZZLE #269 | PUZZLE #270 | PUZZLE #271 | PUZZLE #272 |

237# MEDIUM

238# MEDIUM

60

239# MEDIUM

240# MEDIUM

SOLUTION ON PAGE 67

PUZZLE #265 PUZZLE #266 PUZZLE #267 PUZZLE #268

241# MEDIUM

242# MEDIUM

243# MEDIUM

61

244# MEDIUM

SOLUTION ON PAGE 66

SOLUTION ON PAGE 66

PUZZLE #261

PUZZLE #262

PUZZLE #263

PUZZLE #264

245# MEDIUM

246# MEDIUM

247# MEDIUM

62

248# MEDIUM

SOLUTION ON PAGE 65

SOLUTION ON PAGE 65

PUZZLE #257

PUZZLE #258

PUZZLE #259

PUZZLE #260

249# MEDIUM

250# MEDIUM

63

251# MEDIUM

252# MEDIUM

SOLUTION ON PAGE 64

PUZZLE #253

PUZZLE #254

PUZZLE #255

PUZZLE #256

253# MEDIUM

254# MEDIUM

64

255# MEDIUM

256# MEDIUM

SOLUTION ON PAGE 63

PUZZLE #249

PUZZLE #250

PUZZLE #251

PUZZLE #252

257# MEDIUM

258# MEDIUM

65

259# MEDIUM

260# MEDIUM

SOLUTION ON PAGE 62

PUZZLE #245

PUZZLE #246

PUZZLE #247

PUZZLE #248

261# MEDIUM

262# MEDIUM

263# MEDIUM

66

264# MEDIUM

SOLUTION ON PAGE 61

265# MEDIUM

266# MEDIUM

267# MEDIUM

67

268# MEDIUM

SOLUTION ON PAGE 60

PUZZLE #237

PUZZLE #238

PUZZLE #239

PUZZLE #240

269# MEDIUM

270# MEDIUM

271# MEDIUM

68

272# MEDIUM

SOLUTION ON PAGE 59

PUZZLE #233

PUZZLE #234

PUZZLE #235

PUZZLE #236

273# MEDIUM

274# MEDIUM

275# MEDIUM

69

276# MEDIUM

SOLUTION ON PAGE 58

SOLUTION ON PAGE 58

PUZZLE #229

PUZZLE #230

PUZZLE #231

PUZZLE #232

277# MEDIUM

278# MEDIUM

70

279# MEDIUM

280# MEDIUM

SOLUTION ON PAGE 57

PUZZLE #225

PUZZLE #226

PUZZLE #227

PUZZLE #228

281# HARD

282# HARD

71

283# HARD

284# HARD

SOLUTION ON PAGE 56

| PUZZLE #221 | PUZZLE #222 | PUZZLE #223 | PUZZLE #224 |

285# HARD

286# HARD

287# HARD

72

288# HARD

SOLUTION ON PAGE 55

289# HARD

290# HARD

73

291# HARD

292# HARD

SOLUTION ON PAGE 54

PUZZLE #213

PUZZLE #214

PUZZLE #215

PUZZLE #216

293# HARD

294# HARD

74

295# HARD

296# HARD

SOLUTION ON PAGE 53

PUZZLE #209

PUZZLE #210

PUZZLE #211

PUZZLE #212

297# HARD

298# HARD

75

299# HARD

300# HARD

SOLUTION ON PAGE 52

SOLUTION ON PAGE 52

PUZZLE #205

PUZZLE #206

PUZZLE #207

PUZZLE #208

301# HARD

302# HARD

303# HARD

76

304# HARD

SOLUTION ON PAGE 51

PUZZLE #201

PUZZLE #202

PUZZLE #203

PUZZLE #204

305# HARD

306# HARD

307# HARD

308# HARD

SOLUTION ON PAGE 50

PUZZLE #197

PUZZLE #198

PUZZLE #199

PUZZLE #200

309# HARD

310# HARD

78

311# HARD

312# HARD

SOLUTION ON PAGE 49

PUZZLE #193

PUZZLE #194

PUZZLE #195

PUZZLE #196

313# HARD

314# HARD

315# HARD

316# HARD

SOLUTION ON PAGE 48

PUZZLE #189

PUZZLE #190

PUZZLE #191

PUZZLE #192

317# HARD

318# HARD

319# HARD

320# HARD

SOLUTION ON PAGE 47

PUZZLE #185

PUZZLE #186

PUZZLE #187

PUZZLE #188

321# HARD

322# HARD

323# HARD

81

324# HARD

SOLUTION ON PAGE 46

PUZZLE #181

PUZZLE #182

PUZZLE #183

PUZZLE #184

325# HARD

326# HARD

82

327# HARD

328# HARD

SOLUTION ON PAGE 45

SOLUTION ON PAGE 45

PUZZLE #177

PUZZLE #178

PUZZLE #179

PUZZLE #180

329# HARD

330# HARD

331# HARD

83

332# HARD

SOLUTION ON PAGE 44

PUZZLE #173

PUZZLE #174

PUZZLE #175

PUZZLE #176

333# HARD

334# HARD

84

335# HARD

336# HARD

SOLUTION ON PAGE 43

SOLUTION ON PAGE 43

| PUZZLE #169 | PUZZLE #170 | PUZZLE #171 | PUZZLE #172 |

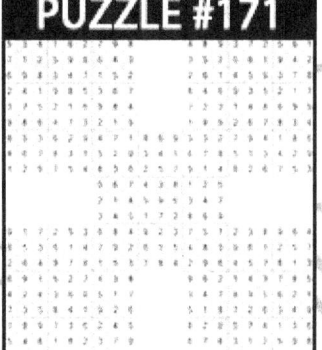

337# HARD

5	3				4	2	6			3	4	8				6	1
		6		8	1					6	4		2				
8	7		2				1		5			7			4	8	
		5	8	2		4					1	5	3				
	2				9			7						9			
7		8			1					3			5		4		
		3				8					4						
3	4		1		2			3		8		5	7				
		3						5									
				3	1	6											
		1		7		4		9									
				8	9	2											
	2						4										
2	6		5		9			7		8		4	1				
	1				4				2								
6		8		3			9			3	8						
	5		6		2			1									
	4	6	3	5			1	2	6	5							
9	2	3		4	3		9	5	2								
	5	4	2		2	8	1										
4	8	6	9	2	4	2	3	6	9								

338# HARD

85

339# HARD

340# HARD

SOLUTION ON PAGE 42

341# HARD

342# HARD

343# HARD

344# HARD

SOLUTION ON PAGE 41

PUZZLE #161

PUZZLE #162

PUZZLE #163

PUZZLE #164

345# HARD

346# HARD

87

347# HARD

348# HARD

SOLUTION ON PAGE 40

PUZZLE #157

PUZZLE #158

PUZZLE #159

PUZZLE #160

349# HARD

350# HARD

351# HARD

352# HARD

SOLUTION ON PAGE 39

PUZZLE #153

PUZZLE #154

PUZZLE #155

PUZZLE #156

353# HARD

354# HARD

355# HARD

89

356# HARD

SOLUTION ON PAGE 38

SOLUTION ON PAGE 38

PUZZLE #149 **PUZZLE #150** **PUZZLE #151** **PUZZLE #152**

357# HARD

358# HARD

359# HARD

90

360# HARD

SOLUTION ON PAGE 37

PUZZLE #145   **PUZZLE #146**   **PUZZLE #147**   **PUZZLE #148**

361# HARD

362# HARD

363# HARD

91

364# HARD

SOLUTION ON PAGE 36

PUZZLE #141   PUZZLE #142   PUZZLE #143   PUZZLE #144

365# HARD

366# HARD

92

367# HARD

368# HARD

SOLUTION ON PAGE 35

PUZZLE #137

PUZZLE #138

PUZZLE #139

PUZZLE #140

369# HARD

370# HARD

371# HARD

93

372# HARD

SOLUTION ON PAGE 34

PUZZLE #133   PUZZLE #134   PUZZLE #135   PUZZLE #136

373# HARD

374# HARD

94

375# HARD

376# HARD

SOLUTION ON PAGE 33

| PUZZLE #129 | PUZZLE #130 | PUZZLE #131 | PUZZLE #132 |

377# HARD

378# HARD

379# HARD

95

380# HARD

SOLUTION ON PAGE 32

SOLUTION ON PAGE 32

PUZZLE #125

PUZZLE #126

PUZZLE #127

PUZZLE #128

381# HARD

382# HARD

383# HARD

96

384# HARD

SOLUTION ON PAGE 31

PUZZLE #121 PUZZLE #122 PUZZLE #123 PUZZLE #124

385# HARD

386# HARD

97

387# HARD

388# HARD

SOLUTION ON PAGE 30

PUZZLE #117

PUZZLE #118

PUZZLE #119

PUZZLE #120

389# HARD

390# HARD

391# HARD

98

392# HARD

SOLUTION ON PAGE 29

SOLUTION ON PAGE 29

PUZZLE #113

PUZZLE #114

PUZZLE #115

PUZZLE #116

393# HARD

394# HARD

395# HARD

99

396# HARD

SOLUTION ON PAGE 28

PUZZLE #109

PUZZLE #110

PUZZLE #111

PUZZLE #112

397# HARD

398# HARD

399# HARD

100

400# HARD

SOLUTION ON PAGE 27

401# HARD

402# HARD

403# HARD

101

404# HARD

SOLUTION ON PAGE 26

PUZZLE #101

PUZZLE #102

PUZZLE #103

PUZZLE #104

405# HARD

406# HARD

407# HARD

102

408# HARD

SOLUTION ON PAGE 25

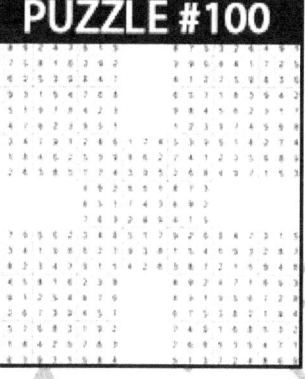

409# HARD

410# HARD

411# HARD

412# HARD

SOLUTION ON PAGE 24

SOLUTION ON PAGE 24

PUZZLE #93

PUZZLE #94

PUZZLE #95

PUZZLE #96

413# HARD

414# HARD

415# HARD

104

416# HARD

SOLUTION ON PAGE 23

PUZZLE #89

PUZZLE #90

PUZZLE #91

PUZZLE #92

417# HARD

418# HARD

419# HARD

420# HARD

SOLUTION ON PAGE 22

PUZZLE #85　　PUZZLE #86　　PUZZLE #87　　PUZZLE #88

421# HARD

422# HARD

423# HARD

424# HARD

SOLUTION ON PAGE 21

PUZZLE #81

PUZZLE #82

PUZZLE #83

PUZZLE #84

425# HARD

426# HARD

427# HARD

428# HARD

SOLUTION ON PAGE 20

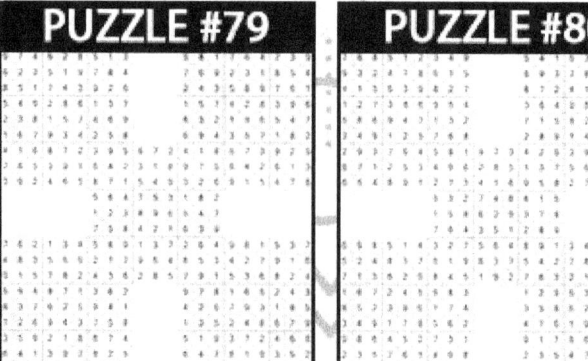

429# HARD

430# HARD

431# HARD

108

432# HARD

SOLUTION ON PAGE 19

PUZZLE #73

PUZZLE #74

PUZZLE #75

PUZZLE #76

433# HARD

434# HARD

435# HARD

436# HARD

SOLUTION ON PAGE 18

| PUZZLE #69 | PUZZLE #70 | PUZZLE #71 | PUZZLE #72 |

437# HARD

438# HARD

439# HARD

440# HARD

SOLUTION ON PAGE 17

PUZZLE #65

PUZZLE #66

PUZZLE #67

PUZZLE #68

441# HARD

442# HARD

443# HARD

111

444# HARD

SOLUTION ON PAGE 16

PUZZLE #61

PUZZLE #62

PUZZLE #63

PUZZLE #64

445# HARD

446# HARD

447# HARD

448# HARD

SOLUTION ON PAGE 15

PUZZLE #57

PUZZLE #58

PUZZLE #59

PUZZLE #60

449# HARD

450# HARD

451# HARD

113

452# HARD

SOLUTION ON PAGE 14

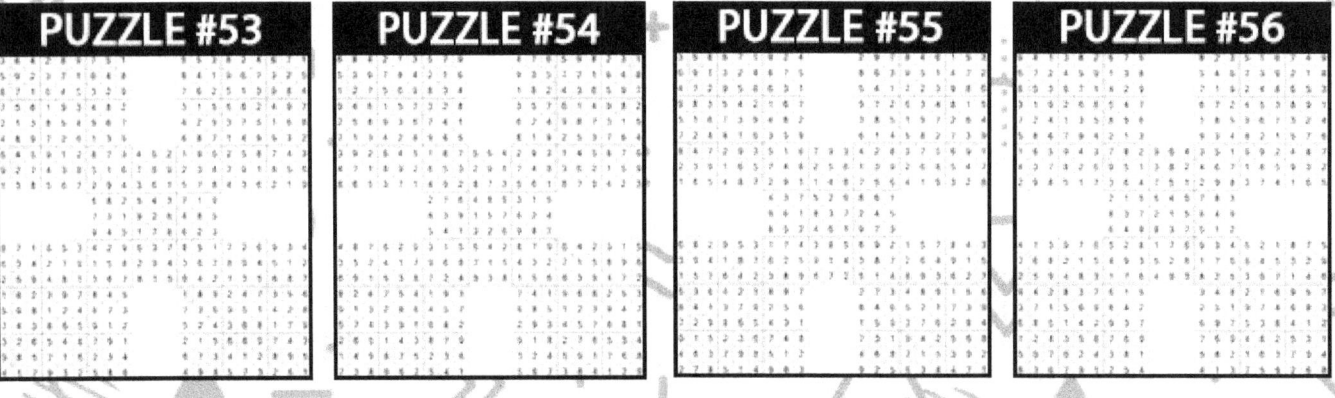

PUZZLE #53   **PUZZLE #54**   **PUZZLE #55**   **PUZZLE #56**

453# HARD

454# HARD

455# HARD

114

456# HARD

SOLUTION ON PAGE 13

SOLUTION ON PAGE 13

PUZZLE #49

PUZZLE #50

PUZZLE #51

PUZZLE #52

457# HARD

458# HARD

459# HARD

460# HARD

SOLUTION ON PAGE 12

| PUZZLE #45 | PUZZLE #46 | PUZZLE #47 | PUZZLE #48 |

461# HARD

462# HARD

463# HARD

464# HARD

SOLUTION ON PAGE 11

PUZZLE #41

PUZZLE #42

PUZZLE #43

PUZZLE #44

465# HARD

466# HARD

467# HARD

468# HARD

SOLUTION ON PAGE 10

PUZZLE #37

PUZZLE #38

PUZZLE #39

PUZZLE #40

469# HARD

470# HARD

471# HARD

118

472# HARD

SOLUTION ON PAGE 9

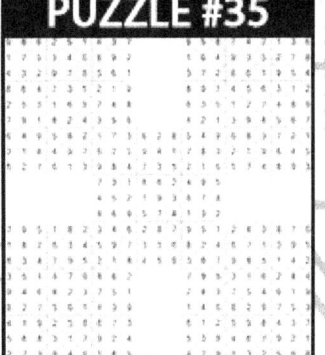

PUZZLE #33 **PUZZLE #34** **PUZZLE #35** **PUZZLE #36**

473# HARD

474# HARD

119

475# HARD

476# HARD

SOLUTION ON PAGE 8

| PUZZLE #29 | PUZZLE #30 | PUZZLE #31 | PUZZLE #32 |

477# HARD

478# HARD

120

479# HARD

480# HARD

SOLUTION ON PAGE 7

PUZZLE #25

PUZZLE #26

PUZZLE #27

PUZZLE #28

481# HARD

482# HARD

483# HARD

121

484# HARD

SOLUTION ON PAGE 6

485# HARD

486# HARD

487# HARD

488# HARD

SOLUTION ON PAGE 5

PUZZLE #17 PUZZLE #18 PUZZLE #19 PUZZLE #20

123

SOLUTION ON PAGE 4

PUZZLE #13

PUZZLE #14

PUZZLE #15

PUZZLE #16

493# HARD

494# HARD

495# HARD

124

496# HARD

SOLUTION ON PAGE 3

| PUZZLE #9 | PUZZLE #10 | PUZZLE #11 | PUZZLE #12 |

497# HARD

498# HARD

499# HARD

125

500# HARD

SOLUTION ON PAGE 2

| PUZZLE #5 | PUZZLE #6 | PUZZLE #7 | PUZZLE #8 |

501# HARD

502# HARD

126

503# HARD

504# HARD

SOLUTION ON PAGE 1

SOLUTION ON PAGE 1

| PUZZLE #1 | PUZZLE #2 | PUZZLE #3 | PUZZLE #4 |